YOUR KNOWLEDGE HAS VALUE

- We will publish your bachelor's and
  master's thesis, essays and papers

- Your own eBook and book -
  sold worldwide in all relevant shops

- Earn money with each sale

Upload your text at www.GRIN.com
and publish for free

Bibliographic information published by the German National Library:

The German National Library lists this publication in the National Bibliography; detailed bibliographic data are available on the Internet at http://dnb.dnb.de .

Imprint:

Copyright © 2015 GRIN Verlag, Open Publishing GmbH
Print and binding: Books on Demand GmbH, Norderstedt Germany
ISBN: 978-3-668-04394-7

This book at GRIN:

http://www.grin.com/en/e-book/304373/overview-of-basic-tools-and-methods-for-genetic-engineering

Manoj Parakhia, R.S. Tomar, V.M. Rathod, B.A. Golakiya

Overview of Basic Tools and Methods for Genetic Engineering

GRIN Publishing

GENETIC ENGINEERING

by

Manoj V. Parakhia. Rukam S. Tomar, Visha M. Rathod and B. A. Golakiya

Index

1. GENE CLONING PROCEDURES

- The recombinant DNA technology involves the cloning of the higher organisms DNA into the prokaryotic organisms so as to allow these lower organisms to produce the specific product coded in the gene of higher organisms.
- The recombinant DNA technology is simply a "cut-paste mechanism" where a single gene is isolated from one organism (cut) and Inserted to another organism (paste) that will produce the new product or character (phenotype) according to the message (genotype) present in the gene.
- Among many enzymes, the most important are the enzymes used for cleaving the DNA, these enzymes are called Restriction Endonucleases (REN).
- These enzymes are the enzymes that have the capacity to break the DNA from phospho-diester bond between the two adjacent nucleotides.

1.1 RESTRICTION ENDONUCLEASES

- REN are the enzymes produced normally by various types of bacteria as a weapon against invading viruses. These enzymes can cleave the viral DNA and therefore they can restrict the further growth of viruses in that bacterium. Due to this character, they are called *Restriction* Endonucleases.
- Their existence was first identified by **Werner Arber** in early 1960s when he was studying on bacteriophages.
- Arber along with two other scientists Daniel Nathans and Hamilton Smith was awarded Nobel Prize in 1978 for their work with restriction enzymes.
- There are basically three types of REN found in nature…
- Type I REN
- Type II REN
- Type III REN
- Among these three types, Type II REN are most commonly used during gene cloning processes. Their cleavage is site specific.
- Their molecular weight is in the range of 20000-100000 Da. They contain two identical subunits.
- These enzymes can cut both the strands of DNA after recognizing specific nucleotide sequence.

Enzyme	Producer Org.	Recognition Sequence	Type of End
Ecorl	*E.coli*	5' A A T T C 3' 3' T T A A G 5'	Cohesive
BamH₁	*B. amyloliquefaciens*	5' G G A T C C3' 3' C C T A G G 5'	Cohesive
HpaI	*H. parainfluenzae*	5' G T T A A C3' 3' C A A T T G5'	Blunt
Saba	*S. aureus*	5' G A T C3' 3' C T A G 5'	Cohesive

- Over 150 specific Endonucleases have been discovered which recognize more than 40 different sequences present on DNA.
- The specific palindromic identified by these enzymes are known as "Restriction sequence" which are palindromic in nature.
- The palindromic sequence when cleaved by REN, will generate two different types of the ends 1) Cohesive/Sticky / Staggered end & 2) blunt end.
- Cohesive ends have got certain, free single stranded nucleotides where as blunt ends do not have them.
- The nomenclature of REN is done according to the bacterial genera, species & strain in which the enzyme is produced. For e.g. ECoR1 is produced by the R1 strain of *E.coli*
- Various examples of REN are given above in the table.
- The 100% of specificity of REN to cut the DNA at only fixed sequence makes it very useful for using in R-DNA technology.
- The success of R-DNA technology is largely dependent on our ability to cut the DNA at specific defined sites.
- In this way REN play central & a key role in R-DNA technology.

1.2 ISOLATION OF DNA TO BE CLONED

- Generally microorganisms are used as the hosts for producing various biomolecules & metabolites of eukaryotic origin.
- Eukaryotic products are produced by the prokaryotic organisms when they have been given the proper DNA to produce the product.
- The DNA comes from a variety of sources.
- The isolation of DNA is very important first step of R DNA technology.
- One gene is a very small part of entire genome for e.g. one gene of *E.coli* = 0.03% of entire genome & one gene of mammalian genomes are approximately 0.03×10^{-3} % of entire genome.
- In this way to isolate the gene from a very large genome specifically in a correct manner is not an easy procedure.
- But the procedure becomes easy when m-RNA & not the DNA for the specific proteinic product is isolated.
- The procedure becomes easy because corresponding m-RNA is found when the protein is synthesized.
- The m-RNA can be isolated during
    1) Hybrid release translation or
    2) Hybrid arrest translation

1.3 C-DNA SYNTHESIS

The stepwise process of preparing c-DNA is explained below :

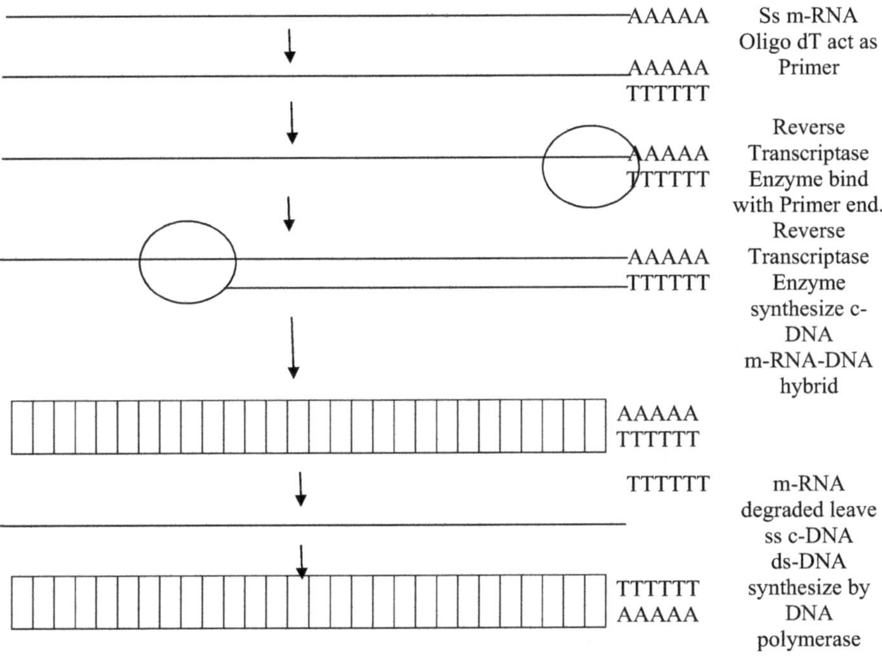

- Reverse transcriptase enzyme has the capacity to catalyze the reaction of preparing DNA as per the sequences of RNA. The entire process will result into the synthesis of complementary DNA (c-DNA) from m-RNA. It is called c-DNA because it is not native DNA but it is produced from the spliced RNA.
- Eukaryotic genes contain non-coding intervening sequence known as "introns" which are removed from the hn RNA (heterogeneous nuclear RNA) after transcription and the coding sequences known as "exons" are joined together to form m-RNA. This process is called splicing of m-RNA.
- When we prepare c-DNA from spliced m-RNA, the c-DNA will not contain introns & so it will be different from native DNA which is a large DNA than c-DNA.
- This feature is very useful in prokaryotic gene cloning because prokaryotes do not have machinery for m-RNA splicing.
- However if one wants to study the native DNA, it would not be feasible to use c-DNA as it does not contain introns. Therefore except this purpose, c-DNA is widely used in R-DNA technology.

1.4 GENE LIBRARY

- In one crude method the entire DNA (genome) of mammalian origin is broken into pieces by multiple REN & all the pieces are cloned into microorganisms & ultimately by using the appropriate selection procedure, desired microorganisms are selected from many unwanted microorganisms. It is called preparation of Gene Library.
- This is a lengthy process because the selection of the desired microorganisms among many undesired microorganisms is time consuming & tedious process.
- Larger the genome & smaller the fragment size, the more colonies have to be grown.
- For e.g. The *E.coli* genome is broken down into 10kb pieces of DNA, total 1.5×10^3 pieces are generated. and if all are cloned in individual organism then 1.5×10^3 colonies are to be grown.
- If mammalian genome is broken into 10kb pieces of DNA, total 2×10^6 pieces are generated and therefore that many colonies are to be grown. This is a very large number and it is almost impossible to handle those and to select a single colony from that.
- Therefore, as a solution to this, the fragment size can be increased so as to decrease the no. of fragments.
- If the fragment size is increased from 10kg to 40 kb then the no. of fragments will be decreased & ultimately 7×10^4 colonies are grown and from them one will be selected
- The upper size limitation of vector to receive DNA is important to determine during the gene library because after DNA is fragment, the individual fragments are inserted in vectors & vectors will not receive the DNA having the size more than 40kg.
- The gene library is an older process but even in present day it is used for some specific gene cloning processes.

1.5 RECENT TECHNIQUE

- Nucleic acid chemistry has advanced today in such a manner it is possible to prepare the desired DNA molecules in the laboratory using modern technology and instruments.
- The process known as PCR technology (Polymerase Chain Reaction) is used to amplify the DNA and to generated DNA in laboratory.
- Even it is possible that we can obtain DNA if we have the proteinic product with us.
- Here we can sequence the protein and generate possible m-RNAs from that. Later on appropriate DNA can be produced from m-RNAs by reverse transcription.
- Such process had been successfully used for the synthesis of the hormone somatostain in bacterial host.
- In this way, the DNA can be isolated by various procedures and used further in recombinant DNA technology to allow the prokaryotes to produce the desired product.

1.6 VARIOUS VECTORS USED IN r-DNA TECHNOLOGY

- Vector is a molecule or organism which carries the DNA from one organism to another during R-DNA technology.
- It should allow the replication of the foreign DNA inserted in it, and moreover it should also allow the selection of an organism having vector by marking some product.
- These are ideal characteristics of a vector and additionally the vector should be able to carry the maximum possible length of the DNA.
- On the basis of these features, following vectors are useful during genetic engineering.
  1) Plasmid as vector
  2) Bacteriophages as vector
  3) Cosmid as vector

1.7 PLASMID CLONING VECTOR

- Plasmids are small extra chromosomal circular DNA molecules which gie additional characteristic to the organism.
- Plasmids are separated from the main chromosomal DNA and they are separately replicated, independent of chromosomal replication (for this, they have their own origin of replication) and pass in the daughter cells during the cell division.
- Some very imp genes are carried by plasmids including antibiotic resistance, sexual factor, colicin, production, antibiotic production, toxin production, Nitrogen fixation, degradation of polymers, production of enzymes, cryptic gene etc.
- All these products are imparting additional features to the organisms that contain the plasmids.
- The presence of replication origins, markers gene and the frequent exchanging of genetic material between plasmids and chromosomal DNA make them very interesting for a biotechnologist who wants a vector during R-DNA Technology.
- Plasmids offer the convenient replication origin but there are many unwanted regions present on plasmids that can be removed when they are used as vector.
- The similar observations were made by two biotechnologists Bolivar and Rodriguez who prepared successful plasmid vector known as pBR^{-322} which is commonly used as vector during R-DNA Technology.
- Besides pBR-3zz (plasmid of Bolivar and Rodriguez), pBR-327 and PUC (plasmid of University of Callifornia) are also used in vector preparation.

1.8 CONSTRUCTION OF pBR^{322}

- pBR-322 is a plasmid prepared in laboratory (*in vitro*) that contain the useful feature of different sources combined together to give a fully functional vectors pBR-322.

- As the requirement of certain characteristics is important in a vector, this plasmid has been prepared to fulfill all of them.
- It is the plasmid from which all unnecessary genes have been deleted & the gene having useful feature are only kept.
- Origin of replication has been taken from a plasmid occurring in *E. coli* known as Col E1.
- pBR-322 contains approximately 20 restriction sites and there are two antibiotic resistance genes in pBR-322 namely amp^R and tet^R (resistance against ampicilin and tetracycline respectively)
- Plasmid replication is not dependent on chromosomal DNA replication as plasmid is self replicating molecule having its own Ori C.
- It is possible to generate more copies of plasmid per bacterial cell by temporarily stopping the replication of main chromosomal DNA.
- Until the chromosomal DNA replication is completed, the cell will not divide but plasmid can replicate in this condition also, so copy no. can be increased.
- Practically copy no up to as much as 3000 plasmids can be generated.
- Due to the presence of marker genes the selection of plasmid is done easily.
- The presence of plasmid in bacterium will confer resistance to ampicillin and tetracycline.
- Furthermore, it is found that few restriction sites are present on the marker genes, so the plasmid DNA can be nicked from there to insert foreign DNA at this site and then the marker DNA is no ore functional.
- After recircularization of plasmid the two broken pieces of marker the gene will flank the foreign DNA.
- This will be the useful feature for differentiating recombinant plasmid (plasmids containing DNA in it) from other normal plasmids. This selection is illustrated in the next page in a diagram.
- The useful features of plasmid pBR 322 make it very useful for using as a vector during R-DNA technology.
- Smaller the size of plasmid, more are the chances of inserting large DNA into it.
- There are two reasons for using the smallest possible plasmid as a vector.
    1. The small molecules of DNA are less easily damaged by shearing during isolation steps. This will allow the violent handling of plasmid during deproteinization and other steps of R-DNA Technology.
    2. The chances of transformation increase as the size of plasmid decreases.

- Small molecules of plasmid are easily taken up by bacterial cells during transformation and large size will not allow it.
- Moreover, due to the smaller and smaller size of the plasmid, length of the foreign DNA can be increased to some extent.
- In this way pBR-322 is a very much useful plasmid cloning vector during R-DNA Technology providing all the necessary functions to a biologist.

1.9 VIRAL DNA / BACTERIOPHAGES AS VECTOR

- Although plasmids are excellent vectors but for larger pieces of foreign DNA, they have a size limitation.
- They cannot be much useful during the preparation of mammalian gene library. When the larger fragments of DNA must be cloned.
- To overcome this problem, the natural gene transfer process – transduction can be used.
- Viruses have the ability to transfer the pieces of bacteria DNA along with them from one individual organism to another individual organism by the process known as transduction.
- Among the variety of bacteriophages, ds DNA containing λ phage is most useful as a vector.
- The entire genome of λ phage is 49 kb and it is contained in protein head.
- The most useful part of the λ phage is one unique 12 nucleotide long complementary ss-DNA sequence known as cos site and it is the key part of λ phage.
- After the recircularization of viral DNA by joining of complementary cos sites, the DNA will be replicated by rolling circle model and concatemer will be generated.
- Ultimately during the "heedful" packaging mechanism, the concatemer will be broken at cos site and DNA will be packed into protein capsid.
- The two important characteristics of λ phage make it useful in gene library.
  1. *The Flexibility of packaging* : The size of the DNA packed in the head can be variable between 79-109% of the total genome. So any size between this two will be packed efficiently.
  2. *Possibility of removing undesirable genes* : A large no of genes can be deleted without affecting the growth and lifecycle of the viruses. It is possible to remove 40% of total λ genome and replace it with foreign DNA.
- On the basis of this, there are two types of phage vectors.
  (a) Insertion vector (b) Replacement vector.
- Insertion vector has only one cleavage site where as the replacement vector has two. Unwanted DNA removed from replacement vector is known as "stuffer" that can be separated from cos site containing left and right arms by centrifugation or velocity gradient ultra centrifugation or electrophoresis as shown in given figure.
- The figure shows the occurrence of various types of the ligation products.
- It is observed that wanted and unwanted fragments of the DNA are inserted between the two arms.
- Only the length that falls in the range of 79-109% of total λ genome will be packed in the artificially supplied heads or natural heads.
- It is also possible that more than one pieces of DNA ligate with each other (multiple ligation) between two arms, this is an unwanted product and will be eliminated during approapriate selection technique.
- It is also possible to have a marker gene in λ phage DNA.

- If the gene for β–galactosidase enzyme is present in stuffer, it will be useful as a marker.
- Lactose negative bacteria when infected with virus will become lac+ve and can hydrolyze the artificial substrate x-gal to give blue coloured product.
- Two types of the plaques will be generated – blue colored plaques and colorless plaques.
- They have two type of viruses :
  (a) Normal λ virus and (b) replacement vector.
- The selection of colorless plaques will allow as to select the replacement vector.
- In this way transducing viruses like λ phage are useful as vector and can transfer the length of DNA up to 21kb.

1.10 COSMIDS AS VECTOR

- Although bacteriophages serve as a good source of vector during R-DNA technology, yet the need for the larger DNA insert remains unfulfilled.
- As we know even in replacement vectors there are many genes responsible for viral growth and multiplication (60% genes of total λ genome)
- These genes are of no use to a biotechnologist who wants to clone the DNA, but this genes will have to be expressed for the sake of vectors.
- As the essential genes make up about 60% of viral genome, the maximum length between two site which can be packed is 52kb and the maximum size of foreign DNA insert is only 21kb.
- When we are using λ phage as vector during mammalian gene library many occasions comes when one wishes to insert even larger DNA than 21kb.
- For this purpose, one unique hybrid vector has been developed which contains cos site of bacteriophages and origin of replication + marker genes of plasmid.
- This unique hybrid vector is known as "cosmid", the word coined by combi9ning cos site of bacteriophages, and plasmid.
- Cosmids are very efficient vectors that does not contain any gene for viral assembly.
- The size of cosmid including a 12 nucleotide long cos site is just 5 kb.
- Linearization of cosmid is done by restriction enzyme and foreign DNA is inserted into cosmids as shown in the figure given below.
- Here also a range of products will be generated but this will include mainly the molecules consisting of foreign DNA with each of its ends joined to a cosmid.
- If the distance between the two cos site falls in the range of 37-52 kb, the packing can be done.
- Here we have to supply packaging enzyme and head and tail proteins from outside in an *in vitro* packaging system.
- As the length of the cosmid is 50kb, the foreign DNA of the size 32-47kb can be packaged in the phage head by the virus of cos site present in the hybrid vector cosmid.

- In the consequent step, the cosmid is efficiently transformed in *E. coli* cells where it will be replicated like a plasmid after recirularizatoin.
- Thus, we can conclude that the cosmids are most efficient vector having the maximum upper size limit of foreign DNA.

1.11 JOINING OF THE DNA WITH VECTOR

- The success of r-DNA technology does not only depend on cutting the DNA with restriction site but also on the appropriate jointing of DNA and vector.
- If the foreign DNA and vector contain complementary cohesive ends, they will be joined very easily if they collide. This joining will be reversible and temporary.
- But it will be made permanent by using the enzyme ligase which will form the phosphodiester bond between 5' phosphate and 3' hydroxyl groups.
- Incubation of samples at lower temperatures for few hours can add up in efficient joining.
- The usual problem observed during joining is recirculation of the plasmid vector.
- This can be avoided by using certain techniques discussed later.
- The two non identical cohesive ends can be converted into blunt end by digestion of ss DNA tail or generation of complementary ss tail.
- These blunt ended DNA molecules can be joined by ligase enzyme, but the bonding will not allow the recovery of the foreign DNA in future because there is no presence of restriction site in the recombinant vector.
- Use of certain techniques allows us to solve two limitations during the joining.
    - (a) Recicularizatoin of plasmid vector,
    - (b) Recovery of foreign DNA from vector.

- To get DNA of first limitation the enzyme alkaline phosphatase is used.
- Alkaline phosphatase enzyme will remove the 5' phosphate molecules from the vector so it will not allow the rejoining of vector as such.
- The vector will only be joined if the foreign DNA containing 5' phosphate is inserted into it, forming recombination vector as shown in diagram.
- Preferably the vector and DNA should contain the similar restriction sites or identical cohesive ends (isoschizomers)
- But it might not be possible that the similar cohesive ends are present in both.
- In such cases any one cohesive ends (either of foreign DNA or of vector) can be converted in to blunt form using the enzyme S_1 nuclease which will digest the single stranded nucleotides.
- Then in the next steps they are attached with DNA linkers which are the DNA fragments containing restriction sites. They are chemically synthesized oligonucleotides in nature.
- There are so many linkers commercially available today so it is possible now to overcome the problem of compatibility and non compatibility in joining of foreign DNA.

- One more method known as Homopolymer Tailing is also available to join the foreign DNA with appropriate vector.
- Here the blunt ended molecules will be treated with the enzyme terminal transfrase.
- This enzyme will insert the nucleotides at the blunt end.
- Here the foreign DNA and vector are treated and dATP and dTTP respectively which are inserted at the end of DNA and vector.
- As poly-A tail will be added at one end and poly-T tail will be added at another end, on mixing, this complementary sequences can join together which can be transformed later on.
- One drawback of this technique, the recovery of the DNA becomes very default.
- It is observed that the above mentioned technique known as Homopolymer tailing, when uses poly-G and poly-C tails, then automatically the site for the restriction enzyme *Pst*-I will be generated.

1.12 TRANSFORMATION AND GROWTH OF CELL

- After preparation of recombinant vector, it must be taken up by the bacterial cell in the process known as transformation.
- Here, the *E.coli* culture which is used must be lacking the restriction system in it.
- This is compulsory because otherwise the *E.coli* will cleave the recombinant vector at restriction sites.
- One more characteristic which is desired in *E.coli* is competence for transformation. Only competent cultures are found to be transforming.
- The treatment of $CaCl_2$ given to the *E.coli* culture growing in exponential phase will increase the chances of transformation.
- The treatment of $CaCl_2$ transformation.
- Moreover, if $CaCl_2$ treatment is given at low temperature, the efficiency of transformation will be increased.
- After the treatment of $CaCl_2$,a mild heat shock or mild electric pulse will result in the uptake of DNA by bacterial cell.
- Generally the efficiency of transformation is increased by applying both the conditions together.
- Approximately 10^5 - 10^7 transformants (*E.coli* who have received the DNA) will be obtained per microgram (µg) of super coiled pBR^{322}. therefore, by transformation we obtain high no. of transformants but this no. will decrease if we want to transform recombinant pBR^{322} and only less than 10^4 transformants are generated per µg of relegated plasmids. This no. decreases progressively as the size of the inserted DNA increases. Linear DNA is completely ineffective in transformation.
- Transformation takes place only with 0.01% of total DNA molecules.
- Thus, the most significant limiting factors for transformation are size as well as shape of DNA which is to be inserted in *E.coli*.
- In this way, the transformation will be completed when the DNA is taken up and appropriately inserted into the *E.coli* chromosome.

- The selection of transformed cells depends on the marker genes. In most cases the marker gene is for the resistance to antibiotic.
- The transformed *E.coli* are first growth on the medium without antibiotics for one hour and then grown on the medium containing antibiotics.
- This is necessary because in the first hour transformed organism will express their antibiotic resistance genes and then only they can resist the antibiotic.

1.13 SELECTION OF CLONES

COLONY HYBRIDIZATION

- The transformed *E.coli* are grown and large n. of colonies will be produced, from this large number we have to select our desired colony that contains the recombinant vector (vector having foreign DNA in it) inserted in them.
- For this purpose, one technique is used that utilize the radio-labeled DNA probe-the technique known as Colony Hybridization.
- Here, replica plating of the plate containing many colonies including of desired one, is done on the nitrocellulose filter and the mirror image of colony pattern is later on grown on fresh agar plate.
- After the immobilization of denatured DNA, it is reacted with the radio labeled DNA probe which contains the sequence complementary to a part of foreign DNA.
- Unbound probe is washed off and the radio image or X Ray image is developed which will reveal the bound probe.
- The probe in the bound state can be directly co related with the presence of the foreign DNA and therefore the selection of the colonies from the original master plate is done on the basis of results of the developed X-Ray image.
- In this way, we can collect the colonies containing the foreign DNA by using radioactive DNA probe.
- The requirement of DNA probe is mandatory during the selection of clones and efficient selection is done if an appropriate probe is used.
- But as the probe itself is a piece of DNA, having complementary sequence to the foreign DNA, it becomes very difficult to get the proper probe.
- For obtaining the probe the technique which is applied is known as Hybrid Release Translation. It is a multi step procedure using in vitro translation system.
- The intention here is first to isolate total RNA and then from it total m-RNA and ultimately after a series of steps the appropriate probe is obtained.
- The entire series of steps for obtaining the DNA probe is explained below.
- The important technique for the identification of appropriate protein product is immuno-precipitation or SDS-PAGE (Sodium dodecyl sulphate – Poly Acrylamide Gel Electrophoresis) which are very much specific in indentifying the desire protein.
- The entire technique involves translation from the hybrid of DNA-RNA and so it is known as Hybrid Release Translation.

- In hybrid release translation the DNA-RNA hybrid is directed for producing the product and then the product is identified by electrophoresis, immunoassay, bioassay, enzyme assay or by biosensors.
- An another related technique, though comparative less useful, is used in few processes and it is known as Hybrid Arrest TRT Translation.
- In this technique the m-RNA which is not bound to the DNA is translated and we are looking for the absence of desired product.
- Nowadays the advanced techniques have made possible to produce the synthetic probes of any DNA sequence.
- If the amino acid sequence is known, then various possible nucleotide sequences can be predicated.
- In this way, nowadays it is easier to get the appropriate probe due to modern technologies.

1.14 EXPRESSION OF CLONED DNA

- After transformation and selection, we have obtained the appropriate host cell which is having foreign DNA inserted in it.
- Now it is most important to make the engineered host produce our desired proteinic product.
- The entire lengthy process performed before this step is of no use if the foreign DNA not expressed within host cell.
- Normally, the host organism will never try to synthesize the foreign protein in it is not required by it.
- In this condition, we have to take deliberate steps so that the foreign gene is expresses.
- There are certain requirements other than the gene to produce a protein product.
- The very first requirement for the initiation of transcription is "promoter sequence".
- The promoter sequence is located upstream at -10 and -35 base pairs.
- We have to provide promoter sequence to begin transcription of foreign gene.
- The product of the transcription is m-RNA which must bind to r-RNA present in ribosome prior to the translation.
- The m-RNA binds with the r-RNA using a sequence known as "Shine Dalgarno sequence" present on it.
- Sine dalgarno sequence is given the name after two Australian scientists. This sequence is complementary to one part of the 16S r-RNA of the ribosome.
- The another requirement for the synthesis of protein is the presence of initiation cadon which must be present on the m-RNA.
- The AUG initiation codon will start the translation of the m-RNA into protein.
- If the c-DNA is used then it will not contain any intron because it is prepared from the m-RNA by reverse transcription.
- But if the native DNA is used during cloning, then it will be containing the intervening DNA sequences known as "introns".

- Introns are present between the exons which are part of the coding DNA and as we have told, introns are part of non coding DNA.
- During translation, introns must be absent and therefore after the transcription the RNA containing introns is spliced. (splicing is the process of removing introns from hn RNA and rejoining exons together to form m-RNA.)
- In r-DNA technology, we must provide splicing machinery if we want to use native eukaryotic DNA which normally contains introns.
- If c-DNA is used then there is no need of using splicing machinery.
- Instead of bothering about all the above requirements, we can provide expression vectors which will be containing all the required sequences.
- The appropriate choice of a promoter is also very important and the promoter must be strong enough.
- In expression vectors normally lac and trp promoters are used which are very much strong and expressible.
- After the proper production of desired protein the problems are not over, we must "protect" the produced protein after the synthesis.

1.15 PROTEIN PROTECTION

- The prokaryotic host cell tends to kill or degrade the eukaryotic protein produced in their cells.
- This can be avoided if the foreign DNA is attached with a short piece of host DNA.
- After the expression of both, there will be a hybrid protein (host protein + foreign protein), this hybrid protein will be identified as "self" by bacteria and they will not degrade it. Therefore indirectly our foreign DNA will also be save.
- There is an alternate mechanism where we attach a gene known as *Pin gene* to the foreign DNA. *Pin* gene is Protease inhibition gene which will produce a product than can nullify the effect of protease enzyme of host bacterium.
- As the protease is inhibited, our foreign protein will not be degraded by the host bacterium.
- The source of *pin* gene is T_4 phages.
- Protein secretion is also a desirable feature during recombinant DNA technology.
- Certain proteins are labeled in such a way that they will be exported out of the host cell.
- If our desired protein is exported of secreted by bacteria then we can save a lot of time and money required in downstream processing.
- Such export of the protein can be done by attaching single polypeptide at its N terminus.
- Thus, protein protection and protein secretion are two essential approaches after the expression of eukaryotic gene in a prokaryotic host cell.
- By providing promoter, Shine-dalgarno sequence, initiation codon, splicing machinery, machinery for protein protection and signal peptide, we shall be able to obtain our desired product in an appropriate manner.

2. GENETIC MANIPULATION OF EUKARYOTIC CELLS

> ### *Limitations of Bacteria :-*

- Normally genetically engineered bacteria are very useful for the large scale production of valuable eukaryotic proteins.
- Bacteria grow rapidly, They grow on relatively simple media, they are easy to handle, they are easily transformed, they can grow on wide range of substrates, they multiply, fast, they contain simple and well characterized genome, they can be easily modified but even though, they have their own problems…

(1) Bacteria do not have introns and so they do not have any system or machinery to remove intorns.
- Therefore that genes containing introns cannot be used in r-DNA technology.
- One look for c-DNA in such cases and it is obtained from m-RNA.
- If the protein is produced in less amount in donor cell then purification of its m-RNA is difficult.
- In such cases. We have to use eukaryotic cells as host cells (Recipient).

(2) Many eukaryotic proteins undergo different types of modifications after they are produced. These modifications are known as "post translational modifications".
- Post translational modification may be a little proteolysis (removed of short polypeptide from protein) or for eg. Conversion of proinsulin into insulin.
- PTM may also be Glycosylation (addition of some oligosaccharides on protein).
- Only eukaryotic cells can carry out such PTM on eukaryotic proteins, bacterial cells cannot do this.

(3) Each bacterial cell usually contains many copies of the plasmid but there is no mechanism to ensure the equal distribution of plasmids in daughter cells.
- Many times during rapid cell division & proliferation some cells appear without any plasmid.
- Such cells without plasmid have selective advantage over other cells having plasmid.
- This happens because plasmid containing cells have to perform some amount of extra work for plasmid replication & gene expression.
- In this condition, they grow slower than cells without plasmid.
- One time comes when the culture is dominated by cells without plasmid and plasmid containing strain is lost from culture.
- To ensure plasmid stability, cells should be grown or medium containing an antibiotic.
- If plasmid has a gene for that antibiotic resistance the only those cells will grow that have plasmid in then.
- This is fine on lab scale but on a large scale, huge amount of antibiotic is deeded which is expensive.

- If plasmid contains one or more genes that promotes-1) bacterial growth rate, they there will be selective pressure for plasmid containing cells.
- For eg. Glutamate dehydrogenase gene inserted in *Methylophilus methylotrophus* which is grown for the production of single cell protein.
- "Mini chromosome vectors which are used in yeast ensures replication and equal segregation of foreign gene in daughter cells.
- Certain plant and animal vectors are also maintained through the generations if they are integrated with nuclear DNA.
- In this way, bacteria have many limitations for their use is gene manipulation and therefore one should look for eukaryotic host cells.
- And moreover, genetic engineering is not only the production of valuable polypeptides in cell cultures many crops are improved by genetic engineering and animal livestock is also improved by the same.
- Therefore some vectors should be developed that can deliver DNA in eukaryotic cells.

2.1. GENETIC MANIPULATION OF PLANT CELLS

> **Ti Plasmid (Tumor Inducing Plasmid)**

- Ti plasmid is considered as the most efficient. Plant vector. It was first time discovered in 1983.
- This plasmid is found in *Agrobacterium tumefaciens*.
- This organism is a typical plant pathogen that enters the dicotyledonous plants from their root system and damages them.
- *Agrobacterium tumefaciens* the fresh wound attach itself to the wall of an intact cell. Later on it transfer a short piece of Ti plasmid into the nucleus of plant cell. This DNA piece is known as T-DNA.
- T-DNA carries few genes which are expressed within the plant Construction of Ti plasmid :-

(1) One gene codes for an enzyme which is responsible for the synthesis of opine from amino acids.

- Opines are not found normally in plants and not metabolized by plants, but they can be used as a nutrient by Agrobacteriaum.
- Total 24 types of opines are detected in nature the particular opine produced by plant depends on the strain of bacterium.
- Some strains produce a gene responsible for the synthesis of "Nopaline" and others have the gene for the synthesis of "Octopine". These genes are respectively known as Nos & ocs.

(2) The other gene found in T-DNA is responsible for the induction of disorganized growth of infected plant cells that forms callus, gall or tumor. It is called one gene.

- Due to this characteristic of tumor production, Ti plasmid is named so (Tumor Inducing plasmid).
- It is also observed that certain Agrobacterial strains contain genes for plant growth hormones Auxin & cytokining tostead of one gene, such hormones also contribute to accelerate plant growth.

➤ *Transformation of Ti plasmid*

- Inside the plant cell; the T-DNA does not remain independent but it becomes integrated into plant chromosome.
- The integration occurs due to the presence of two repeated sequences of 25 base pairs which are located near the ends of T-DNA.
- Once Ti plasmid is nicked, T-DNA is released from and it travels into plant cell.
- T-DNA becomes circular in its intermediate form, and ultimately it is inserted into plant chromosome DNA.
- Genes that remain on Ti plasmid include those for attachment of bacterium to plant cell wall, those for transfer of T-DNA and those for uptake and catabolism of opine.
- The gene which is responsible for the transfer of T-DNA lies on Ti plasmid and it is called vir-Region.
- This transformation is irreversible and callus cells can continuously produce opine once they are infected (One type of opine) is also present on Ti plasmid and it is called noc gene.

➤ <u>*Ti Plasmid as a Vector :*</u>

- Usefulness of Ti plasmid as a vector during plant genetic manipulation is due to many features of Ti-plasmid these features include…
- (1) The genes are T-DNA is eukaryotic in nature. Even though these genes are of bacterial origin, they are transcribed by plant RNA polymerase. These genes contain introns and they are removed correctly during m-RNA splicing.
- (2) Only some part of T-DNA is required if one wishes to use it as a vector.
- Opine production genes and tumor formation gens can be replaced by foreign DNA.
- Thus nos/ocs & onc genes can be deleted.
- If nos or ocs genes are removed, their promoters are retained to ensure the gene expression of foreign DNA.
- Such experiments have been successfully carried out for the expression of kanamycin & methotrexate resistance in plants.
- (3) Even one gene can be used as a marker DNA which allows cells to grow at higher rate without the addition of hormones from outside.
- But if one wants to grow healthy plant, their one gene should not be included and should be deleted.
- In this way, Ti plasmid is excellent as a vector during the genetic manipulation of plant cells.

- ➤ ***Ti Plasmid and promoters :***
- As the Ti plasmid is very large in size, approximately up to 235 kilo base pairs, it is not modified as a whole.
- Instead, all the modifications & manipulations are done on T-DNA and then normal T-DNA is replaced with modified T-DNA in a Ti plasmid.
- It is not always feasible to use opine promoters as they do not ensure selective expression of our foreign DNA.
- This occurs because opine promoters are always active for the continuous production of opine by plant.
- Opine structural genes are removed but when opine promoters are retained, they are permanently active.
- Therefore, it is desirable to have regulated promote instead of opine promoters.
- One of such regulated promoter is phaseolin promote.
- Gene for bean phase Olin has been expressed in sunflower. This gene was under the control of phase promoter.
- This promoter sequence can be isolated and attaches with any gene that is to be transferred in plant cell.
- This regulated promoter is very useful for gene many.
- Phaseolin is synthesized only in developing seeds. There of it seems that its promoters are regulated one.

- ➤ ***Limitations of Ti-Plasmid***
- Most plants of commercial interest are belonging monocotyledonous plants and agrobacterium cannot induce tumor in them.
- Most of our major crop plants, cereals, grains etc. are monocots but Ti plasmid cannot be used as vector for desire genetic manipulation in them.
- Although Ti plasmid can be used for most dicotyledond plants, it has limitation for monocots.
- However, as every black cloud has silver living we are optimistic about the use of Ti plasmid for monocots.
- It is now experimentally proved that *Agrobacterium tumefaciens* transfer its T-DNA into some monocots resulting in expression of opine gene but without inducing tumor formation.

- ➤ ***Cauliflower Mosaic Virus :***
  Cauliflower Mosaic virus (CaMV) is a useful vector for manipulating plant cells.
- The useful feature of CaMv is that the naked viral DNA is infective. It can enter plant cells directly are it is rubbed onto a leaf with mild abrasive.
- After entering the cells, DNA is replicated, expressed and encapsulated within virus particles.
- Newly packed viruses invade the rest of the plant.

- CaMV DNA is not integrated in host genome therefore it may not be handed over to all progeny cells, but rapid spread of viruses throughout the plant fulfills our requirements.
- The Limitations: of CaMv include the following…
  - i)   Size of CaMv genome can nm is increased much.
  - ii)   Almost all genes of CaMv genome are essential aspect should not be deleted. Therefore there is less room for foreign DNA insert.
  - iii) CaMv has a narrow host range.

Above listed limitations make CaMv unlikely to be used as vector.

> ### *Direct transformation :*

- In this technique, plant protoplasts are prepared by treating plant cells with poly ethylene Glycol (PEG).
- Protoplasts can take up DNA from their surrounding medium.
- These uptakes DNA can be stably integrated into plant chromosomal DNA.
- This technique has been used to transform plant cells with a gene for kanamycin resistance, linked to a strong promoter.
- Plants regenerated from this plant were resistant to kanamycin, and this resistance was inheritable.
- The limitation of direct transformation is that the uptake DNA is randomly integrated in plant genome.
- As long as one can select the desired plant among all undesirables, this method should not present a great problem.

2.2 GENETIC MANIPULATION IN MAMMALIAN CELL.

- It is extremely difficult to culture mammalian cells in vitro.
- Mammalian cells do not have totipotency, so what genetic changes we wish to do in vivo, we should do at early embryonic stage.
- Mammalian growth rate is very low and the no. progenies produces by there is also very limited.
- Due to all these reasons, genetic manipulation of mammalian cells is limited to the production of hormones, monocloned antibodies and viral particle (for vaccine) by manipulating and culturing man cells.
- Moreover, due to their high cost of production, the prices of such products remain too high.

(1) Direct transformation:

- The foreign DNA is precipitated with calcium and this mixture is treated with host mammalian cells.
- The hosts cells take up DNA and once inside, the DNA molecular are ligated to give concatemer.
- The concatemer is there integrated as one large blockin nuclear DNA.

- Concatemer formation is useful as a selectable marker gene can be inserted in between.
- A marker gene that allow the cells to use xanthine as a precursor for the synthesis of purines, instead of hypoxanthine, is useful.
- Another marker gene gives resistance to neomycin which normally inhibits protein synthesis.

(2) Viruses:
- Viruses like SV40 is useful as vector in mammal gene cloning.
- Viruses, not only delivers DNA, but they also ensure DNA replication and contain promoters which are useful for the gene expression of foreign DNA.
- SV40 (simian virus) contains a circular DNA molecule of 5.2 kb length.
- This DNA contains….
  (1) Replicating origin.
  (2) Early genes – For DNA Replication.
  (3) Late genes – For Viral Coat Proteins.

- The foreign DNA can replace either 'early' or 'late' genes, but then we have to co infect host cells with 'helper' virus which contains functional copies of missing genes.
- If cell line known as COs is used as host, there is no need to use a helper virus to supply missing genes.
- In the cells of COs cell line; the missing genes are integrated into the nuclear DNA.
- Use of COs cell line overcomes the problem of separating recombinant and helper viruses from each other after their recovery from the cells.

(3) Microinjection:
- The cell lines can not generate whole animals as they lack totipotency.
- For the breeding and improvement of whole animal, DNA must be transferred into the nucleus of fertilized egg using microinjection.
- The injected DNA randomly enters into the chromosomal DNA.
- If it is attached to a suitable promoter, it might be expressed.
- Site of insertion has a main influence on the success of method. Results are unpredictable.
- After microinjection, the egg must be re-implanted in a surrogate mother.
- This is a slow and labor intensive method of genetic manipulation.
- This method, nevertheless, has been used to transfer the gene of rat growth hormone in mice, resulting in the production of 'giant' mice.

2.3 GENETIC MANIPULATION IN YEAST:

- Plant and animal cells have very slow growth rates and are very difficult to cultivate in laboratory environments.
- Due to the above reason, bacterial cultures are used as host cells, but they also have their limitations.
- Bacteria do not have introns and so do not have machinery to remove them (Splicing Machinery).
- They are not able to perform post translational modifications.
- They do not assure equal distribution of plasmid vectors in next generation's cells.
- Even if they are not pathogenic, they can exchange genes with pathogens and become pathogenic.

> ***Advantages of yeast.***

- Almost all the limitations of plant cell animal cell and bacteria can be overcome by the yeast *Saccharomyces cerevisiae*.

(1) It is eukaryotic with a small, well characterized genome.

(2) It has a very much faster growth rate then animal and plant cells.

(3) It is non pathogenic and do not exchange DNA with any pathogen.

(4) Many of yeast genes contain introns, which are correctly spliced out,
However the introns themselves contains sequence for their own removal (self splicing), so it is necessary to with c-DNA as the foreign DNA.

(5) Yeast can also carry out "post translational modifications "Like proteolysis, glycosylation etc.

(6) Stability of cloned DNA is higher in yeast as it contain mini-chromosomes to which foreign DNA should be attached. Mini chromosomes are equally distributed in progeny cells.

(7)

3. BLOTTING TECHNIQUES

1. Analysis of DNA by Southern blotting.
2. Analysis of RNA by Northern blot Hybridization.
3. Analysis of protein by western blot techniques.
4. Do blot & slot blots.

3.1. ANALYSIS OF DNA BY SOUTHERN BLOTTING

> In 1975 E.M. southern give a procedure to identify the location of genes and other DNA sequences on restriction fragment separated by gel electrophoresis. That's why called Southern blotting.
> This technique include following steps.
> In first step, the sample of DNA is converted into fragment by the treatment of multiple restriction enzymes.

- In second step, the electrophoresis of the DNA fragment is carried out, but at that time the denatiwation of DNA is also carried cut by alkaline treatment.
- Then after the band of DNA fragment percent on the gel is transferred on the nitrogen cellular paper.
- Then after the fragment is immobilized on the fiter by drying or UV induced cross linking.
- In next step the radio liable probe, which complementary to our DNA fragment is added on the filter paper & this probe is Hybridized or annealed with owl desired DNA fragment & double helix DNA fragment is formed.
- Then After the unbound probe is washed etc.
- In next step for the detection of our desired fragment X-Ray is exposed to the Nitrogen cellulose paper & the X-Ray film is developed by autoradiography.

❖ *Application*
- This techniques is very useful for the analysis of DNA that is isolated from two different organism.
- Analysis of RNA by Northern blot Hybridizations.
- Similar to DNA analysis, the analysis of RNA is also carried out by northeru blotting techniques.
- This is so called northean blotting because it is the mirror image of southern blotting lechniques.
- In this technique the fragmentation is carried out by RNA are enzyme.
- Then after the fragment of RNA is denatured by Formaldehyde or other chemical denaturant.
- In next step the electrophoresis of this RNA fragment is carried out & the band of RNA fragment is transferred on the Nitrogen cellulose paper.
- Then after the Radio able DNA or RNA probe is added to the nitro cellulose paper for the detection of our desired RNA fragment.

❖ *Preclusion*
- Fragmentation of RNA by RNA are is very sensitive. There for must be worn a gloves during analysis to prevent the contamination of solution with RNA are present on our finger.
- The glass-ware that are used in analysis must be free from the RNA ase, which done by proper chemical treatment because it is very stable & can not denatwled by bailing or other treatment.

❖ *Application*
- This techniques is used in the analysis of gene expression, we can determined whether the gene is expressed in all tissue or only in certain tissue of expression of individual genes during growth & development.

❖ *Disadvantage*
- Major disadvantage of this tech is that it only measured the accumulation of RNA transcripts.

3.2 ANALYSIS OF PROTEIN BY WESTERN BLOT TECHNIQUES.

- Poly acryl amide gel electrophoresis bas been used as important tools for the separation & characterization of protein.
- As we know that protein is composed of two or more poly peptide chain & the electrophoresis of such protein is carried out in presence of "sodium dodsyl sulfate" (SDS) which denature the protein.

- ***Detection***
- After electrophoresis, protein is detected on gel by treating with coomassive blue or silver stain.
- However some times polypeptide in gel can also be transferred to nitrogen cellulose paper & in visual protein detected by using specific antibody.
- This transfer of protein from gel to nitrogen cellulose paper is called "western blotting" & is performed by using an electric current also called "electro blotting
- Non bounded antibody are then washed & then bound antibody can be detected by adding secondary antibody.
- This secondary antibody is labeled with radioisotope or by enzyme which give color product with chromogenic substance.

- ***Application :-***
- This process is very powerful tool for indentifying & characterizing specific gene product.

- ***Do Blots & Slot Blots :-***
- Another variety of southern or northern blots is dot blots & slot blots.
- Instead of performing southern blotting, unknown DNA can be identify by dot blots or slot blots.
- Here the spot of difffernet DNA sample is placed on nitro-cellulose paper.
- Then After DNA is denatured & immobilized and ultimately radioactive probe against our desired DNA is added on paper.
- It only bind with our desired DNA & we can detect the location of our desired DNA from many spot. This is called dot blot techniques.

4. DNA SEQUENCING

- This answer include following points
  (A) Introduction.
  (B) DNA sequencing by chemical modification.
  (C) DNA sequencing by chain termination Method
  (D) DNA sequencing by PCR

- DNA sequencing is very useful for the early of genetic engineering.
- Following factor help to developed the method for sequencing of DNA.
    a) Discovery of restriction enzyme.
    b) Improvement of gel electrophoresis.
    c) Development of gene cloning give large quantities of particular gene.

4.1 DNA SEQUENCING BY CHEMICAL DEGRACATION METHOD.

- This techniques was first developed by Maxam & Gilbert.
- Determination of DNA sequence is depend on the resolution power of gel electrophoresis.
- Poly acrylmide gel electrophoresis give separate band of single stranded DNA which are only differ in bases.
- In this chemical degradation method first the 5' ro 3' end is labled with p.
- Then after the DNA is partially modified with chemical reagent specific for different base and cleaved at the modified nucleotides.
- This will generate a set of molecules which is different in length but same in isotapically labeled terminal.
- Then after the gel electrophoresis of the fragment is carried out.
- Then after the sequence of DNA is determined by the result of gel electrophoresis by Auto radiography.

4.2 SEQUENCING BY CHAIN TERMINATION

- This technique was developed by Sanger.
- In practice chemical degradation method is replaced by chain tearmination method.
- In this procedure DNA is synthesized rather than degraded, but principle is similar to chemical method.
- Following step are followed for Sanger's method.
- Add primary, a small oligo-nucleotides sequence added to DNA sample.
- Then after the DNA sample is divided into four tube.
- Add DNA polymerase & all four Nucleotide tri-phosphate in each tube.
- Add ddGTP, ddATP, ddTTP, ddCTP in different four tube repetitively.
- Electrophoresis is all these four tube is carried out on same gel by preparing four well in gel.
- Band of DNA fragment is detected by Autoradiography because ddNTP is labeled with 32p isotope.
- Then after the sequencing is carried out.

4.3 DIRECT DNA SEQUENCING BY USING PCR.

- ➢ The polymerase chain reaction is not only used for amplification of DNA but also for sequencing of this DNA.
- ➢ This method of DNA sequencing is faster & more reliable 2 utilize for whole genome DNA or cloned DNA fragment.
- ➢ The sequencing using PCR involve two step.
    (1) Generation of sequencing template (ds or ss) using PCR.
    (2) Sequencing of DNA by thermo labile DNA polymerase or thermo stable DNA polymerase.
- ➢ Thus, the DNA sequencing method by using PCR, does not need the cloning of DNA in ss DNA phage vector.
- ➢ It is difficult to sequenced the double stranded DNA product of PCR, because DNA strand is re associated after denaturation
- ➢ To reduce this problem, either a variant of the standard method for sequencing ds DNA is employed or by sequencing single stranded DNA template by sequenceing PCR.
- ➢ A number of thermo labile DNA polymerase have been used for sequencing of in vitro amplified DHA, but thermo stable DNA polymerase can also be used for sequencing reaction.
- ➢ This techniques involve the principle of Sanger's chain termination method, in which the incorporation of ddNTP is carried out for chain termination.
- ➢ However maxam & Gilbert's chemical degradation method can also be used.
- ➢ In Sanger's method, as usual for mixtures are prepared, each using one of the four ddNTP.
- ➢ Then after the ^{32}p labeled primer, & the mixture with amplified DNA, Tag polymerase & appropriate buffer are incubated at 70°C for 5 min.
- ➢ The reaction is stopped by addition of form amide stop solution in all four tube.
- ➢ Then after all mixture, are allow to run of poly acrylamide gel. Which give the band.
- ➢ This band can be detected by computer or manually.
- ➢ This method using PCR helped in the automation of DNA sequencing.

YOUR KNOWLEDGE HAS VALUE

- We will publish your bachelor's and master's thesis, essays and papers

- Your own eBook and book - sold worldwide in all relevant shops

- Earn money with each sale

Upload your text at www.GRIN.com and publish for free